The Question & Answer Book

AMAZING WORLD OF PLANTS

AMAZING WORLD OF PLANTS

By Elizabeth Marcus
Illustrated by Patti Boyd

Troll Associates

Library of Congress Cataloging in Publication Data

Marcus, Elizabeth.
   Amazing world of plants.

   (Question and answer book)
   Summary: Answers questions about the different kinds of plants, their growth and reproduction, and the ways they are beneficial to mankind.
      1. Plants—Miscellanea—Juvenile literature.
2. Botany—Miscellanea—Juvenile literature.
[1. Plants. 2. Botany. 3. Questions and answers]
I. Boyd, Patti, ill. II. Title. III. Series.
QK49.M33   1984   581   83-4836
ISBN 0-89375-967-8
ISBN 0-89375-968-6 (pbk.)

Copyright © 1984 by Troll Associates, Mahwah, New Jersey

All rights reserved. No part of this book may be used or reproduced in any manner whatsoever without written permission from the publisher.

Printed in the United States of America
10   9   8   7   6   5   4   3   2   1

What are they?

Some of them are the oldest and the largest living things on earth. Without them, there would be no food to eat and no air to breathe. What are they?

They are *plants*.

How do plants help us?

Plants are a very important part of our lives. Houses, furniture, tools, and toys are made from plants. So are paper, clothing, and medicines. Plants shelter wildlife. Plants protect and enrich the soil. They provide energy to run our factories and light our homes. And most important of all, if there were no plants, there could be no life on earth.

We usually think of plants as being green. But not all plants are green. A plant can be a white birch, a red rose, an orange fruit, a yellow vegetable, and even a brown mushroom!

Where do plants grow?

Plants grow wherever there is the right amount of sunlight, food, and water. Some plants grow best in high sunny fields. Others live in shady forests or on high, rocky mountain cliffs. Some plants live in the oceans, while others live in the deserts. Some plants thrive in damp, tropical jungles. And some even survive in the freezing polar regions.

How many kinds of plants are there?

Botanists, scientists who study plants, say there are over 350,000 different kinds of plants on earth. They divide these plants into two main groups—plants that make seeds and plants that do not make seeds.

Which plants are seed-makers?

More than half of the plants in the world are *seed-makers*. All seed-making plants are either flowering plants or conifers. *Conifers* are plants that produce cones instead of flowers.

What are flowering plants?

Flowering plants include daisies and beans, rhododendrons and forsythias, cherry trees and apple trees, to name only a few. Every flowering plant has the same basic parts. These parts are roots, stems, leaves, and flowers.

What do the parts of a plant do?

The roots hold the plant in the ground. Roots also soak up water and minerals from the soil. Stems carry the water and minerals to the leaves. The stems also hold the plant up to the light. The leaves use the energy from sunlight to change water and minerals into food.

The flowers are the reproductive parts of the plant. They make the fruit that holds the seeds. All flowering plants have fruit of some kind.

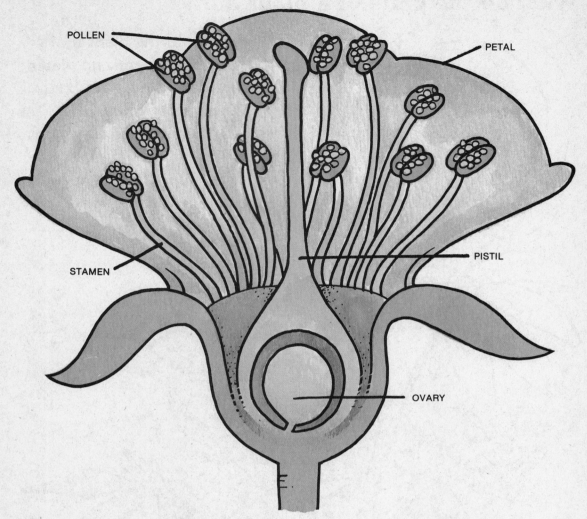

If you look closely at a flower, you will see that it has several different parts. The petals of some flowers look like brightly colored leaves. They shelter the parts of the flower that help the plant reproduce.

The pistil is a tube or stalk at the middle of the flower. Surrounding the pistil are other stalks called stamens. At the top of each stamen is a yellowish dust called pollen. Pollen must get from a stamen to the pistil before a flowering plant can reproduce. This is called *pollination*. If pollen from one flower pollinates a flower on another plant, it is called *cross-pollination*.

How does pollination take place?

Flowers can't pollinate all by themselves. They need help. Sometimes the wind blows tiny grains of pollen through the air. If a pollen grain lands on a pistil, a seed will develop.

Sometimes insects or birds help to pollinate flowers. Flowers produce a sweet liquid, called nectar, that bees and butterflies like. When an insect collects nectar, pollen may stick to tiny hairs on the insect's body or legs. Some of the pollen may brush off on the pistil of that flower, or on the pistil of another flower the insect lands upon.

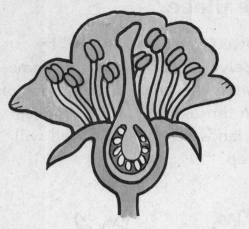

After pollination, seeds develop inside the ovary, which is hidden at the bottom of the pistil. The ovary grows larger and becomes the fruit of the plant. The fertilized seeds are inside the fruit. The fruit may be an apple, a tomato, a pod of peas, or even an acorn. If a seed is planted in the ground, it grows into a new flowering plant.

What are conifers?

Conifers are plants that grow their seeds inside cones. Pine, spruce, and fir trees are some familiar kinds of conifers. Most conifers have two kinds of cones on the same plant. Some cones produce pollen, while others produce seeds. The wind carries the pollen to the seeds. After pollination, the seeds fall to the ground, where they begin growing into new plants.

How many kinds of conifers are there?

There are about 500 kinds of conifers. Sometimes they are called evergreens, because many of them have long, needle-like leaves that stay green all year. Conifers are the oldest and largest living things in the world. Some bristlecone pines in California may have started growing 5,000 years ago. And California's giant sequoia trees are almost 350 feet (105 meters) high and 30 feet (9 meters) wide!

What are the "lower plants"?

Seed-bearing conifers and flowering plants are what most people think of when they hear the word "plant." But there are other important members of the plant family. Some are large. Some are small. None of them have seeds. These members of the plant family are sometimes called *lower plants.* They include ferns, mosses, algae, and fungi.

What are ferns?

Ferns are a form of plant life that has been on the earth for a very long time. Scientists have found the fossil remains of ferns that lived millions of years ago. They think today's seed-making plants developed from the ferns of long, long ago.

Ferns have roots, stems, and leaves—like seed-making plants. But ferns do not grow from seeds. Instead, they have a reproductive cycle that has two stages.

In the first stage, a fern plant produces *spores*. Spores grow in small, round clusters on the underside of the fern's leaves. The clusters open and out come the spores. They are light enough to be carried by the wind. Some spores take root and sprout as new plants within a few weeks. Others may take as long as fifteen or twenty years to sprout!

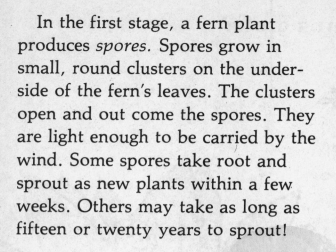

SPORES

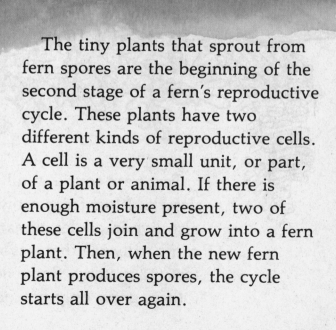

The tiny plants that sprout from fern spores are the beginning of the second stage of a fern's reproductive cycle. These plants have two different kinds of reproductive cells. A cell is a very small unit, or part, of a plant or animal. If there is enough moisture present, two of these cells join and grow into a fern plant. Then, when the new fern plant produces spores, the cycle starts all over again.

Where do ferns grow?

Ferns are found in all parts of the world except in the deserts. Most ferns grow in the warm tropics, but many also live in cooler climates. Some are so small that you can hardly see them, while others grow as high as 40 feet (12 meters). Some act like vines by climbing trees. Others grow only on rocks.

What are mosses?

Have you ever seen spongy tufts of greenery growing in the woods? The tufts are another lower plant form called *mosses*.

Most mosses are very small. The largest known kind, found in Australia, is only 2 feet (.6 meter) tall. Mosses reproduce in two ways. Sometimes part of the plant breaks off and becomes a new plant. Sometimes mosses reproduce by spores. When the clusters of spores burst, the spores explode into the air. They may travel for miles, carried by the wind, before they settle to the ground and start to grow.

Mosses can endure very high and very low temperatures. Sometimes they turn brown and look dead in dry weather, but after a heavy rain they turn green again.

There are thousands of kinds of mosses. They are seldom hurt by insects or disease and can grow almost everywhere except in salt water.

What are algae?

If you have ever been to the seashore, you may have seen another form of lower plant life. Long pieces of seaweed cling to the rocks at the water's edge, or wash up onto the beach, where they dry out. Seaweeds belong to a group of simple plants called *algae*.

Where do algae grow?

Algae are one of the world's oldest groups of plants. Most algae live in water. Seaweed lives in salt water, but other kinds of algae live in fresh water. Some algae live in the soil. And the greenish stains you may have seen on damp rocks and trees are also algae.

TYPES OF ALGAE

PLANT CELL DIVISION

The most common way algae reproduce is by cell division. In cell division, a cell divides, or breaks apart, and each part begins to grow. A new plant eventually grows from each part of the original cell.

21

What are fungi?

Fungi are another group of lower plants. Most fungi reproduce by cell division. They grow just about anywhere—on land, in water, and even in your kitchen! Scientists have counted more than 75,000 kinds of fungi. If you have ever seen mushrooms, mold on bread, or mildew on damp clothing, you have seen fungi.

How are fungi different?

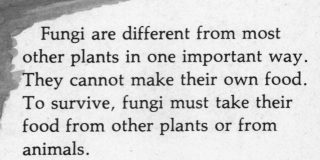

Fungi are different from most other plants in one important way. They cannot make their own food. To survive, fungi must take their food from other plants or from animals.

How do plants produce their own food?

Most plants can produce all their own food because they contain *chlorophyll*. Chlorophyll is a substance that makes the leaves of most plants green. Green plants use chlorophyll to capture sunlight. Then they use the energy from sunlight to make food, in a process called *photosynthesis*.

What is photosynthesis?

Photo means "light" and *synthesis* means "putting together." During photosynthesis, minerals and water from the soil are combined with carbon dioxide from the air. The food that is produced is then sent from the leaves through the stems to all parts of the plant. Some is used right away by the plant, and some is stored.

How does photosynthesis help us?

Something else besides food production happens during photosynthesis. Oxygen is released into the air. People and animals need oxygen in order to survive. When we breathe air, we take in oxygen and release carbon dioxide. Plants take in carbon dioxide and release oxygen. This partnership between plants and animals of all kinds is just one part of the *cycle of nature*.

What is the cycle of nature?

The cycle of nature starts with the sun. Green plants use the sun's energy to change water, minerals, and carbon dioxide into food and oxygen. The oxygen is breathed by animals. The food helps the plants grow.

Many plants become food for animals. The plant-eating animals become food for meat-eating animals. Then, when plants and animals die, they decay and return minerals to the soil. This makes the soil richer, so more plants can grow.

What would happen if there were no green plants?

First, there would be no food for the plant-eating animals, so they would die. Then there would be no food for the meat-eating animals, so they would die. And even if the animals could survive without food, there would be no oxygen to breathe. The cycle of nature would be broken, and the earth would become a lifeless planet.

What else do plants give us?

Long before there were animals and people on the earth, there were plants. Over millions of years, countless numbers of ferns and other plants lived and died. When their leaves, stems, and roots decayed, they enriched the soil.

In some places, thick layers of mosses collected in swamps and formed a material called *peat*. Peat is partly decayed plant matter. It can hold great quantities of water, so farmers and gardeners often plow it into dry soil to make their crops grow better. In some areas, peat is an important source of energy. People cut bales of peat and allow them to dry. Then they burn the peat as fuel.

In many parts of the world, thick layers of peat were buried beneath sand, rock, and soil millions of years ago. Heavy pressure and chemicals in the earth slowly changed the peat into coal. Today, this "fossil fuel" is an important source of energy. When we burn coal, we are really using the energy that was stored by plants that lived millions of years ago.

What is "food energy"?

We also get energy from plants that are living today. This is called food energy. Food energy is stored in almost every part of a plant.

When we eat potatoes, carrots, and beets, we are eating the roots of different kinds of plants. Celery and asparagus are the stems. Lettuce and cabbage are leaves. Apples, peaches, and pears are fruit. And rice, wheat, and corn are seeds. When we eat the roots, stems, leaves, fruit, or seeds of plants, we are taking in food energy. Food energy is the fuel that makes your body work.

Are there harmful plants?

Many growing plants give us food to eat and air to breathe. Most dead plants enrich the soil or give us fuel to burn. But some plants are not so helpful.

Weeds are troublesome plants. If they are not controlled, weeds can crowd out farm crops and vegetable and flower gardens.

Algae can form an unhealthy scum on water.

Tiny grains of pollen in the air can cause diseases like hay fever and asthma.

Many fungi cause plant diseases, such as Dutch elm disease, chestnut blight, and brown rot. Other fungi can cause skin problems, like athlete's foot. Certain kinds of mushrooms are poisonous. And mildew can spoil food and ruin clothing.

Do the benefits outweigh the harm?

Yes. Harmful plants are far outnumbered by those that help people. Lumber from trees is used to build houses and furniture. Wood is also made into paper and cardboard.

The bark of the cork tree is used to make bottle stoppers and fishing floats. The sap of the sugar-maple tree gives us maple syrup. Rubber trees give us rubber for countless products.

The soft fibers attached to the seeds of the cotton plant are made into thread and woven into cloth. You might be wearing clothes right now that were made from cotton.

Plants give us the raw materials to make many other useful products. These products include turpentine, rope and twine, photographic film, fertilizers, and medicines.

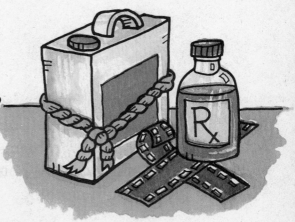

Plants also give us their beauty throughout the year.
New buds and flowers greet us in the spring.
In summer, tall trees shade us from the hot sun.
Colorful leaves brighten the brisk days of autumn.
And the quiet beauty of evergreens stays with us
through the long winter.

How are plants a part of our lives?

Plants are a part of our lives in so many ways. Whenever we are hungry, we eat plants or foods that originally came from plants. And every time we take a breath, we use oxygen that was released by green plants. In return, every time we exhale, we give plants the carbon dioxide they need.

It's a cycle of give and take, borrow and return. And the cycle continues, as it has for countless generations. Plants live and grow, die and decay. Then new life sprouts from the fertile soil...to start all over again.